Island Phoenix

Island Phoenix

A Pictorial Documentary of Life on Kwajalein Island during World War II

Robert S. Fletcher
and
Jonathan S. Fletcher

St Maximus Scriptorium

Published in the United States
by
St. Maximus Scriptorium
Hopkins, SC
www.stmaxscriptorium.com

ISBN-13: 978-0-9838399-9-6
Printed in the United States of America

To all those who worked, lived, and died on Kwaj
so that the rest of us might work, live, and die in freedom.

Contents

Preface

This look at Kwajalein Island is basically a replication of a pictorial documentation of what was going on during the time that my father, Robert Fletcher, was stationed there approximately from January 1944 to January 1946. After the island was captured from the Japanese on February 5, 1944, the U.S. Army Air Corps, the predecessor to our modern U.S. Air Force, moved in to establish a base that was used as a way-station for the injured being evacuated from the front lines of fighting to the west. I really have little record of his duties other than the creation of this document, which he passed along to me pretty much as soon as I was old enough to recognize and appreciate its value—probably when I was in high school.

After having written a few things myself, I finally decided to convert the document to a form that could be read by others who might have an interest in life on an island in the Pacific during WW II. The major work involved was to scan the pages, crop the pictures and retype the text. A pdf file of all the original scans is available for anyone who is interested from the St. Maximus Scriptorium website.

Of course, my own interest mainly arises from the fact that my Dad's imprint is everywhere in the document. Each page has his literary signature, and all the text was written by him. He definitely had a bit of a flare for language. Having been educated at Bowdoin College, he had a liberal arts kind of appreciation of the English language. Another interesting aspect of the document is what he chose to record. I think its great value comes from its almost homespun content. He was interested in the mundane as well as the heroic. Somehow his great appreciation of the people involved in the war effort shows itself at every turn.

So, I would like to say that this modest effort on my part is a tribute to my Dad and all the people he loved and respected who paid a high price—often the ultimate price—for our freedom. He was a collector of swords and guns, but oddly enough, he never used them in any way. I recall that he took a 10-gauge shot gun out once and came back with a badly bruised shoulder from the kick. I think he went hunting one other time. That was it. He wasn't a hunter or a target shooter—he was the quintessential appreciator. I think he saw behind the artifact, the drama of those who used the weapons in some struggle. He seemed to appreciate the heroic nature of humanity. At some point late in life, my mother told me that she had thought that he had been drafted into service. Eventually she found out that he had actually enlisted. I am sure he saw what was going on, and like so many other patriotic Americans, he needed to get involved.

When he was dying of pancreatic cancer in February of 1974, I asked him how he felt about his situation, he said without hesitation, "Well, I saw a lot of guys die in the War—and if they could do it, so can I." To me he was a hero—kind, gentle, thoughtful, courageous, and a man of great integrity. If I have any integrity at all, it is grounded in my fortunate relationship with my father. I have some other things that he did, such as a collection of arrowheads, but I think this documentation of life on Kwajalein Island captures the man better than anything else. I hope any readers might not only get an interesting picture of life on Kwaj, but also get a sense of one man who served his country—and his family— with devotion.

Jonathan Fletcher

July 4, 2023

1
Introduction

Background

The Phoenix was a mythical bird that burned up in a wild conflagration only to rise again from the ashes. My Dad had to have dreamed this title up. That is way above my pay grade. But the image is stunning. As the pictures on the cover indicated, the island was decimate as a result of the battle that took place to secure the island from its previous occupants. I am told that there was only one palm tree standing at the end of the battle. Out of those ashes arose one of the most important military links to forward operations in the Pacific war. As you will see, Aircraft from Hawaii routinely stopped at Kwajalein Island for refueling and arming for the westward journey. Coming back, the planes brought the wounded for a brief respite before continuing on to the Hawaiian Islands for more concentrated care.

Life on Kwaj represented the full scope of human activities from concentrated military responsibilities to a wide range of opportunities to relax between duties. Some of the folks you will see may be relatives. If so, please let me know via email at

jonathanfletcher@stmaxscriptorium.com

The authoritative description of the battle of Kwajalein that precede**d** the Army Air Corps occupation is a book entitled *Island Victory: the Battle of Kwajalein Atoll* by Lt. Col. S. L. A. Marshall (1944)[1] A more up to date reference to Kwajalein is found in *Coral and Concrete: Remembering Kwajalein Atoll between Japan, America, and the Marshall Islands* by Greg Dvorak.[2] Finally there is an amazing video clip on the battle on the History Channel.[3]

Robert Fletcher was born in Portland Maine in 1903. He enlisted in the Army Air Corps in 1944 and served until discharged in January of 1946. His son, Jonathan, that would be me, was born nine months later. You figure that one out.

Robert's last assignment was as information officer on Kwajalein Island, the major island in the Kwajalein Atoll, in the Marshall Group, One of his major duties was to supervise the photographing and annotating of all the various activities taking place on the island after it was secured by the Seventh Infantry Division in February of 1944. Jonathan ended up with the photo album after it had been declassified. This is that record.

---

[1] The most recent edition is by Bison Books, the University of Nebraska Press, 2001, 147 pp. It has many pictures related to the battle in the back of the book.
[2] University of Hawaii Press, 2018, 313 pp.
[3] The following video clip is of considerable value, if one wants to appreciate what the battle was all about: http://www.history.com/topics/world-war-ii/battle-of-kwajalein

This is not a history or a commentary. It is simply an amazing pictorial record of military life on an island in the Pacific that was responsible for some crucial activities supporting the war effort as it moved closer and closer to its ultimate Pacific objectives. The picture captions were all my Dad's. He earned a B.A. degree in psychology from Bowdoin College in Maine. It is this liberal arts background that lent its patina to his writing. In addition to his wonderful image of the Island Phoenix, I can hear his lyrical voice in many of the captions.

Other than whacking all this together, my contribution is minimal. I suppose for me, this is a tribute to a wonderful father, one who felt that serving in the time of war was his duty. Mom told me that she thought he was drafted. In fact, he enlisted. Would that my own sense of duty could rise so high.

In addition, this certainly is a modest tribute to all the men and women pictured here who selflessly served their country in need. While clearly there were friend and foe, this is basically a story of life and dedication.

An Important Note on the Idea of "The Enemy"

The pictures of war are, to say the least, unsettling. We have fought and killed out brother and sister Native Americans, British, Germans and Japanese. Killing is nothing less than a tragedy on all sides. Most who fight are simply supporting their native lands. The pictures shown here are an integral part of that tragedy. They have been retained because they are part of the story as it was told. To avoid them for some sense of political correctness would be to injure the integrity of the works as it was recorded. But we need to be perfectly clear. We all pay a high price for war and for the peace that we hope results. To move beyond enmity to deep abiding friendship is the name of the game. Hate is not a long-term option. Maybe it is not even a short-term option. Let us proclaim our deep friendship with all those who may appear on these pages as a part of a past conflict. If you see some of these pictures and weep, shed your tears for all those whose lives have been lost in the great tragedy of war.

2
The Cost to Japan

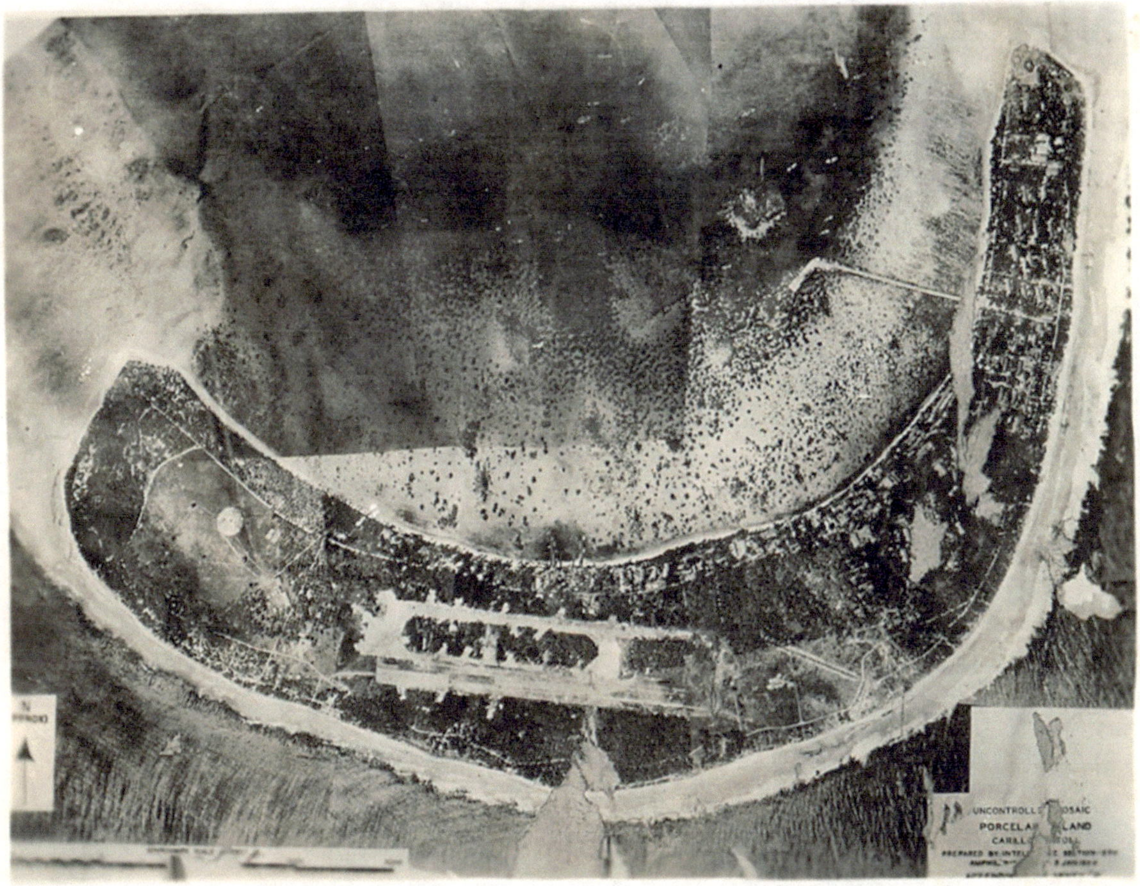

On 5 January 1944, Kwajalein Island (Code–Porcelain) lay more or less peacefully at the southeast corner of the Kwajalein Atoll (Code–Carillon). Much of it was covered with trees and other tropical growth; a small air strip was under construction, and an anti-tank ditch lay diagonally off its east end. Single and dual mount heavy AA [antiaircraft] guns with medium and light AA were emplaced at the northern and western points and along the ocean-shore. The lagoon beach was dotted with pill boxes for small-arms fire. The greater bulk of buildings, barracks, and storage facilities were concentrated on the eastern half of the boomerang-shaped island.

The section at and north of Nob Pier was one of the most heavily settled.

After receiving the “Spruance Haircut” (Naval bombardment under Admiral Raymond A. Spruance), and the “Mitscher Shampoo” (Aerial bombardment under Vice Admiral Marc A. Mitscher), it looked like this from the air.

and like this from the ground.

After the battle, hundreds (See HRR for Jan). [Not at all sure what this abbreviation refers to]

and thousands of Jap dead were gathered up.

and buried in these burial sites;

Their tanks rest on the junk pile;

and their landing craft lie rotting in the lagoon.

Their vaunted heavy dual-mount dual-purpose guns rust in the surf,

and the few remaining shells have been cut up for souvenir ash trays.

A huge power unit lies on its side, kicked there by a titan.

and the last remaining pill box, heavily constructed of reinforced concrete and coral,

stares emptily at the glittering lagoon.

Nob Pier itself came through unscratched, but it's one defensive fortification (center) was wrecked.

Other concrete installations were reduced to rubble, and the few which remained somewhat intact were put to good use.

This one now serves as the Island Guard House.

These steel tanks are so punctured with shell holes they resemble a pair of Swiss Cheeses.

Sub-surface installations fill with rainwater.

And raised circular emplacements resemble clay pigeons at a Tennessee turkey-shoot.

The might that was Japan's now broken in defeat, slowly succumbs to encroaching Nature and will soon disappear from view.

The Japanese left behind not only a key air base when they departed to the bosom of their forebears; from the wreckage here have been retrieved several items of rolling stock which are worthy of passing note. This Nip truck, for example, has the world-shattering distinction of possessing a Chevrolet motor and a Ford-type rear end. It takes 32.6-inch tires, the same size as used by our forces, and its distributor cap, points, and rotor are interchangeable with our Delco-Remy parts.

This right-hand-drive staff car, curiously enough, is a 1937 Graham. According to reliable authority (Walter Riehm, MoMM1c, Ordnance Garage), Japan bought the dies for its construction from the U.S. in 1938. **Somewhat bemused, the recorder of this history [my Dad] sits—**

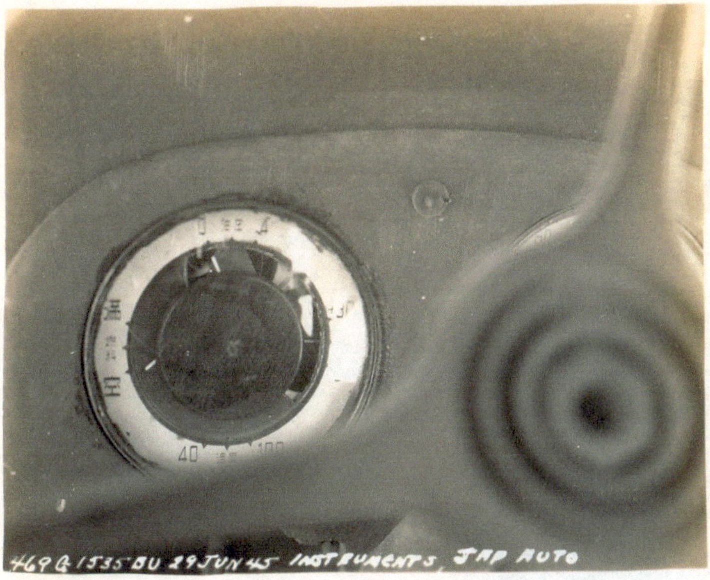

Trying to penetrate the mystery of a Jap gage.

Jap jeep, a half pint corn popper comparable to the Austin—built to accommodate five passengers!

But the last word is always ours. This grinning GI has built himself a scooter out of parts and particles from the junk pile. Low slung and equipped with streamline chassis, head light and double-reversible handlebars, it whips him around the island rapidly and gets approximately 7496 miles to the gallon.

Kwajalein Island lies in the Pacific Ocean at 167º 44´ East 08º 43´ North. Looking NORTH to Ebeye Island.

EAST,

SOUTH,

WEST to Enubuj Island.

[Kwajalein after it was developed by the U.S. Army Air Corps. Compare the length of the airfield and the other facilities to those in the picture of the Island before the battle.]

3
On the Flight Line

From across the far ocean reaches come the D-54s of the Air Transport Command, to spot with unerring accuracy this small way station where they land for rest and revitalization. Presented herewith in pictures is the story of the aircraft served at the 1535th AAF [Army Air Force] Base Unit.

The "spotter" guides the giant plane by arm signals to its assigned parking place, and immediately the servicing and refueling procedure goes into high gear.

Mechanics roll their work stands into position and the passenger stairway is moved up on a finger-lift.

The gasoline truck backs into position for refueling.

And the inspector goes over the "squawk sheets" with the ground crew chief, laying out the work to be done by the ground crew.

Engine cowlings come off in jig time and

In a matter of minutes, the engine servicing and inspection is proceeding under a full head of steam.

While this is going on, radio equipment is checked on board the plane and any discrepancies are corrected in the Communications Shack where

Competent mechanics dig out and exterminate the bugs.

Work proceeds apace on the engines and

Upon completion, they are "run up" by ground crew in a final check.

In the Engineering Shack

the Engineering Officer notifies Operations that the aircraft is ready to go.

The Flight Engineer signs his acceptance of the serviced plane

and in the Operations Building

The pilot is cleared for the next leg of his journey.

A line Quonset hut has been taken over by Priorities & Traffic for use as a Passenger Waiting Room.

The walls are decorated with war maps. The In-Flight Lunch section has a counter, and wooden benches offer relaxation—of a sort.

Crew, passengers and cargo aboard, the four great engines roar once more, and

the plane takes off out over the Pacific wastes once again.

4
Air Evacuation Mission

The evacuation of casualties by air from forward areas to general hospitals both in the Pacific Islands and the United States is the third part of the three-fold Air Transport Command Mission. Other phases have been covered in previous issues (Quartering and messing of transient personnel: March; Servicing of transient aircraft: April).

Tied in with present air evacuation procedures at this station is the 22nd Station Hospital, whose commanding officer is Maj. L. L. Grzesk [pronounced Gresk], shown here conferring with his adjutant, Capt. Hoffman.

The hospital, many of whose buildings still retain the revetments built over a year ago,

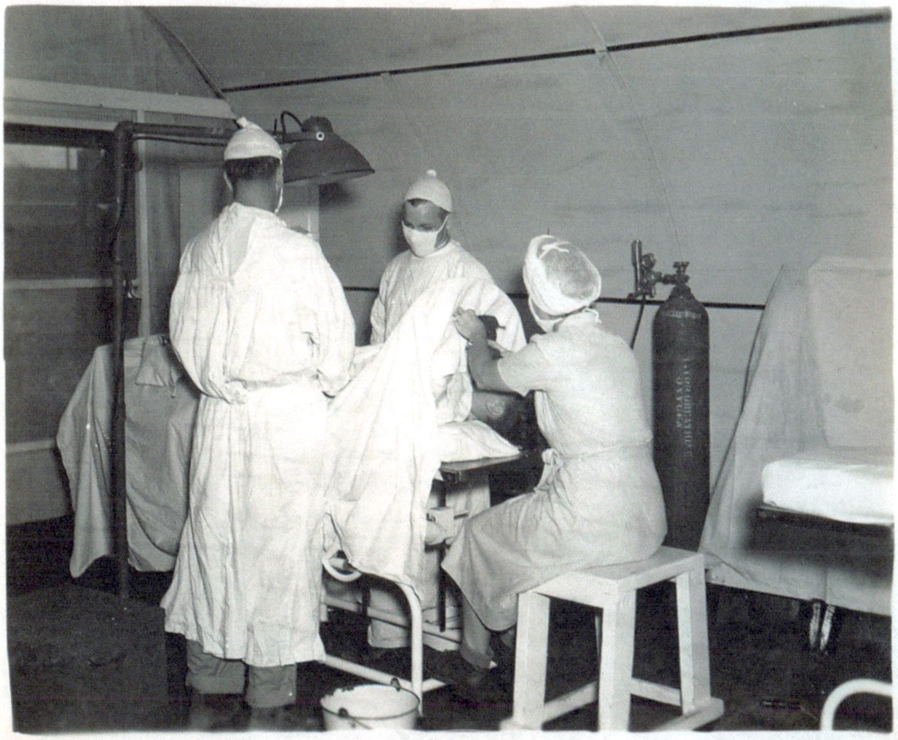

makes all its facilities—operating room,

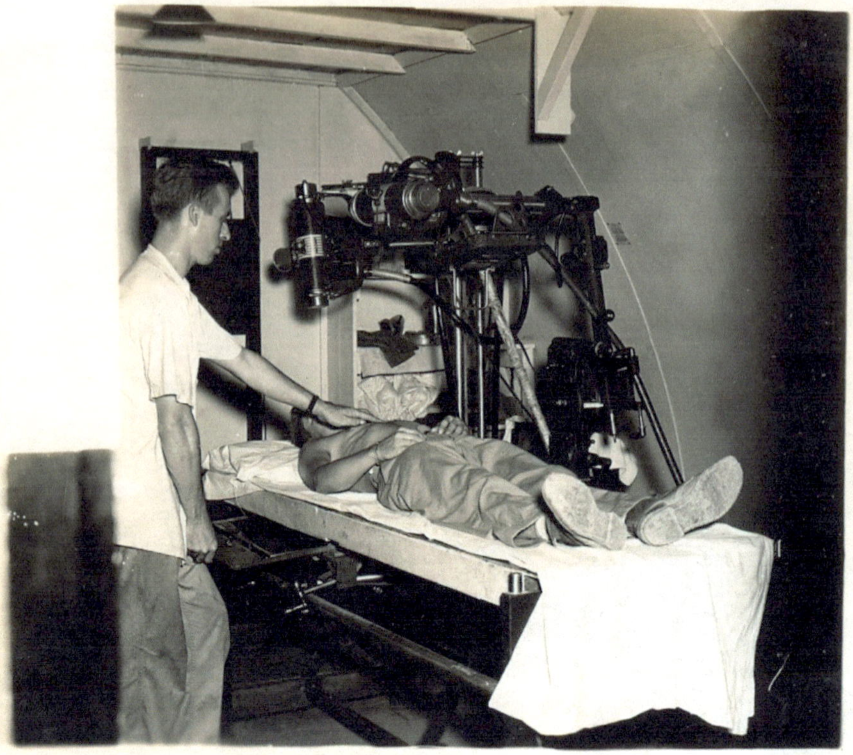

X-ray laboratory,

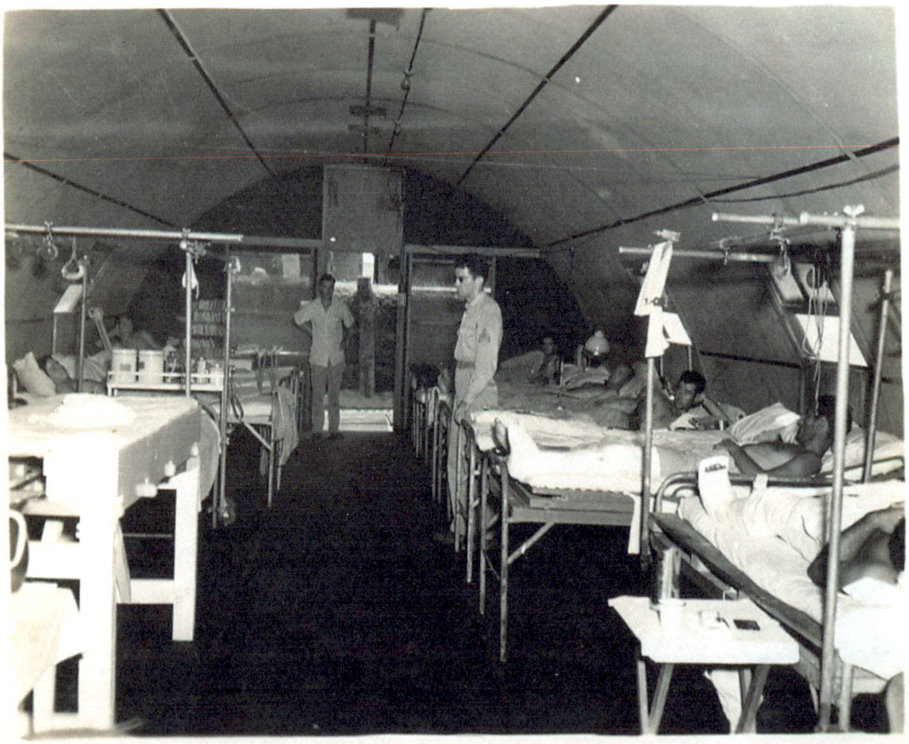

and wards available to casualties from forward areas off loaded here for treatment and hospitalization pending air transport to either Oahu, Hawaii or California.

Air Evacuation Procedure at Kwajalein deals largely with transient patients who stop at this waystation for only an hour or so and then continue their flight eastward. Here, however, is presented a pictorial narrative of a group of litter and ambulatory patients who, having been collected here from neighboring islands and from the battle fronts, are treated, checked, and placed on a hospital plane for evacuation to hospitals better able to provide the necessary treatment.

As the litter patients were being taken from the ward to the waiting ambulance, we stopped and chatted with one. He is S/Sgt Fred Morris of Atlasburg, Pa. In the army 38 months and overseas 22, he was with the 123rd Infantry Regiment, 33rd Division, on Luzon Island when, at 1800 hours on 31 March [1945] he received a Jap machine gun bullet in the knee while going to the aid of two wounded comrades.

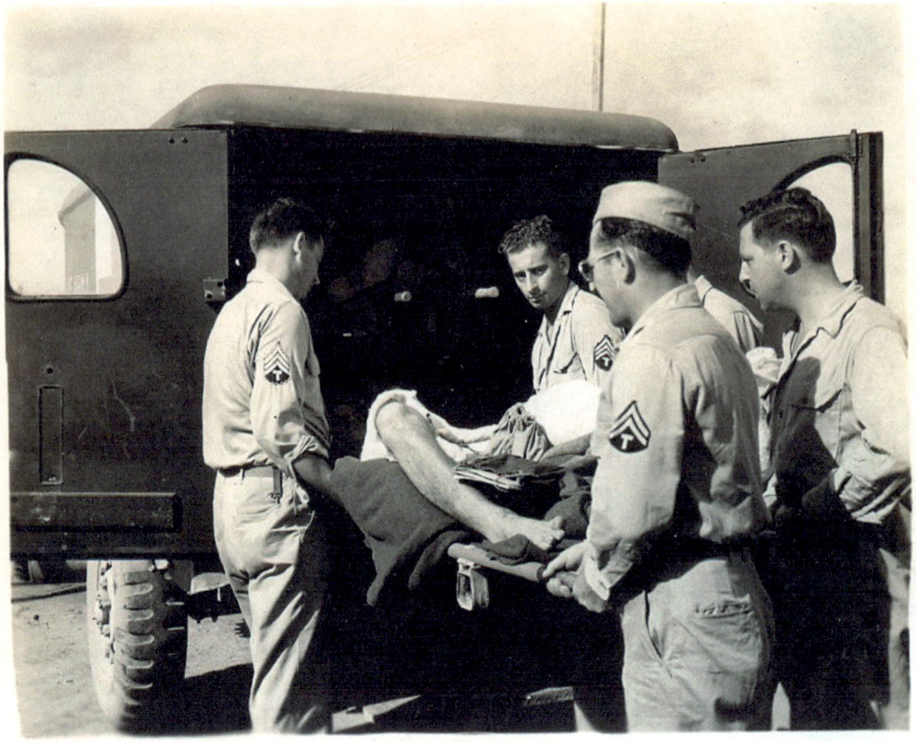

These corpsmen need no "Handle with Care" signs. They ease the litter gently in to keep from jarring the Sgt's wired knee.

Loaded, the ambulance driver carefully checks his passengers before closing up.

The convoy rolls slowly down Ocean Road

and along the taxiway to the waiting Airplane.

Loading of the patients' baggage is completed.

Walking or ambulatory patients go to the nearby "Trade Winds," newly completed American Red Cross rest hut, where ice cream, magazines, and music are provided by the Gray Ladies.

Tastefully painted, decorated with humorous works of art, and furnished with comfortable upholstered chairs and settees, the atmosphere is one of quiet rest, good cheer, and helpful friendliness.

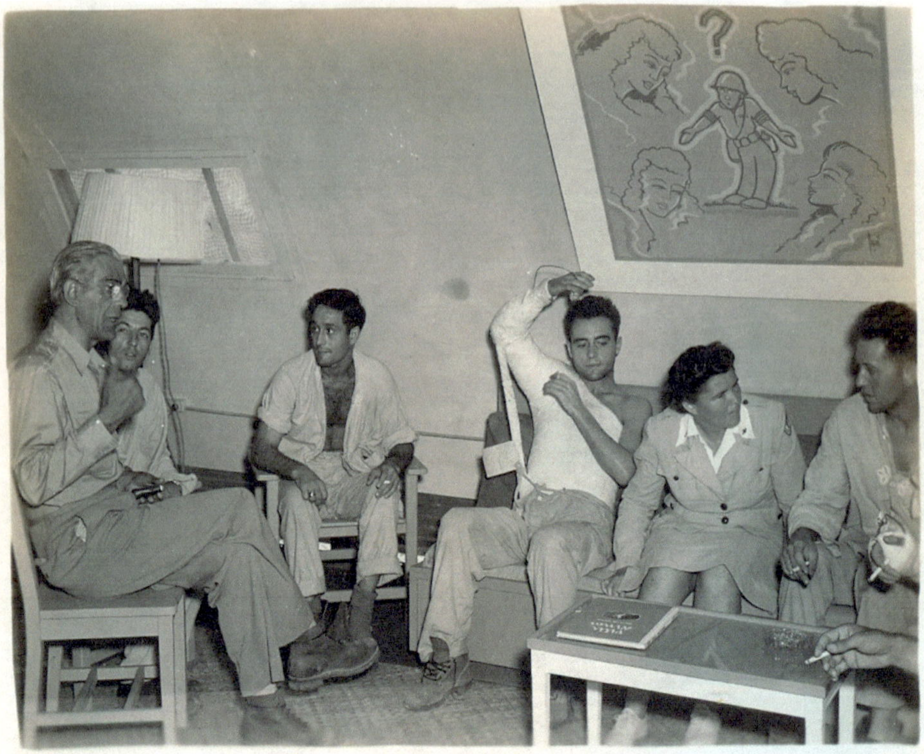

A distinguished visitor, Mr. Boris Karloff, chats with Miss Smith, ARC [American Red Cross] and a group of patients.

Back on the line, Sgt. Morris takes it easy while another litter patient is carried to the forklift.

Going UP!

Morris's turn comes,

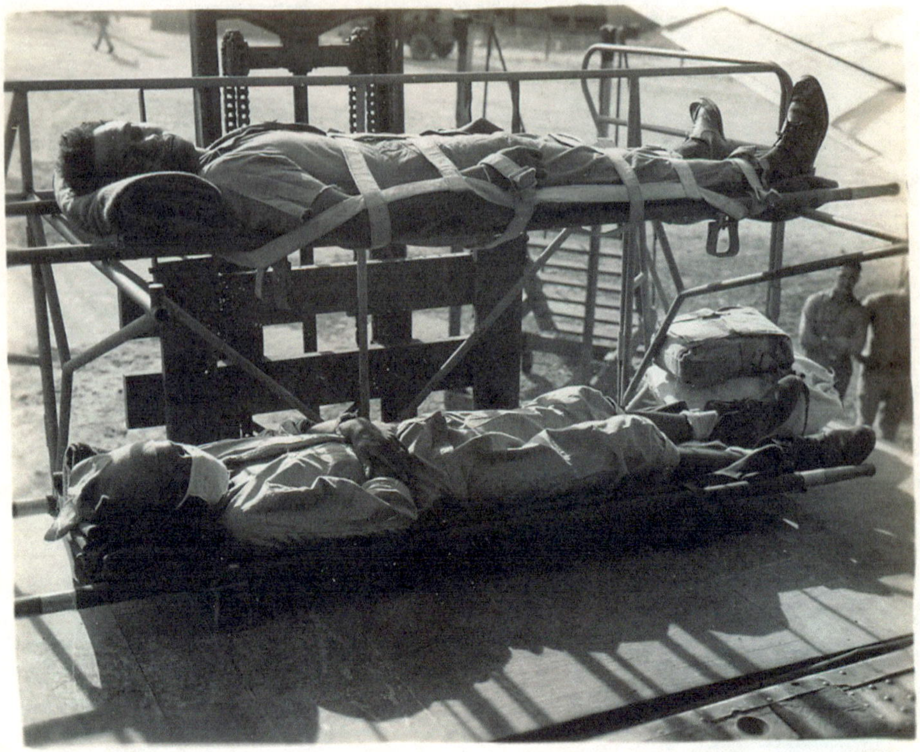

and he is followed by two other soldiers, one a psychiatric case [top].

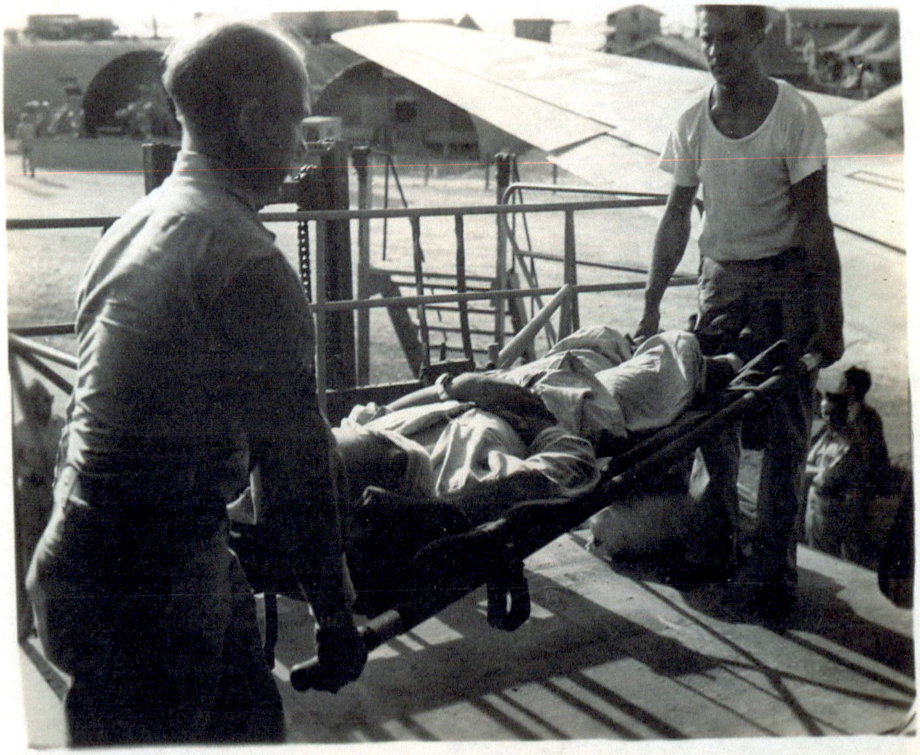

Removed from the lift, the litters are secured to the walls of the fuselage, each patient being carefully located according to injury and need for accessibility. Patients with minor injuries are seated along the length of the plane opposite the slung litters.

The Air Evacuation Surgeon checks medical records with the Flight Nurse who has boarded the hospital aircraft with her complete kit and equipment. An expert in her line, she can administer oxygen, morphine, blood plasma, can redress wounds, and give medical and surgical attention wherever needed during the coming flight.

With all patients aboard and final preparations completed, the forklift has removed the passenger stairway, and the Flight Traffic Clerk closes the door.

An ambulance driver watches the hospital plane roar off the runway on the next leg of its important journey.

5
Combined Operations Building

On 28 April 1945, Operations and Communications left their small Quonset hut, Navigation & Briefing abandoned its wooden shack, and Base Weather vacated its offices. These activities assembled under one roof in this Nissen hut, whose sides had been opened to provide light and ventilation, and whose interior had been carefully planned and partitioned.

Flight Operations occupies a large section to the left of the main entrance,

Communications just beyond,

and Navigation and Briefing has set up its charts, photos, and displays across the corridor.

The 53rd AAF [Army Air Field] Weather Station occupies an inner section,

where it operates its Main Forecasting Room

and observing section with map plotting table.

In addition to these offices, Weather operates a Radiosonde Station in a hut next to ARC [?].

In a neighboring building weather balloons are filled with hydrogen

and sent aloft to be observed.

6
Oceanic Air Traffic Control

A key function in the operation of trans-Pacific flights, Oceanic Air Traffic Control is well housed in a Quonset hut east of the Control tower and next to the Motor Pool (left).

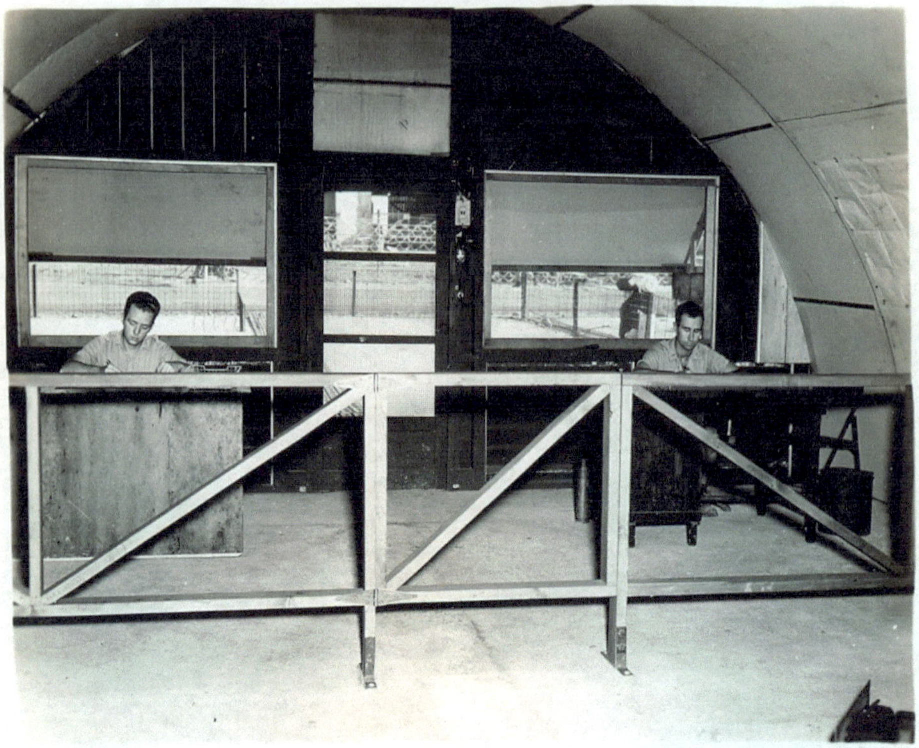

The officer-in-charge and the chief civilian technician occupy the front section of the building, the remainder of which is given over to

operating personnel

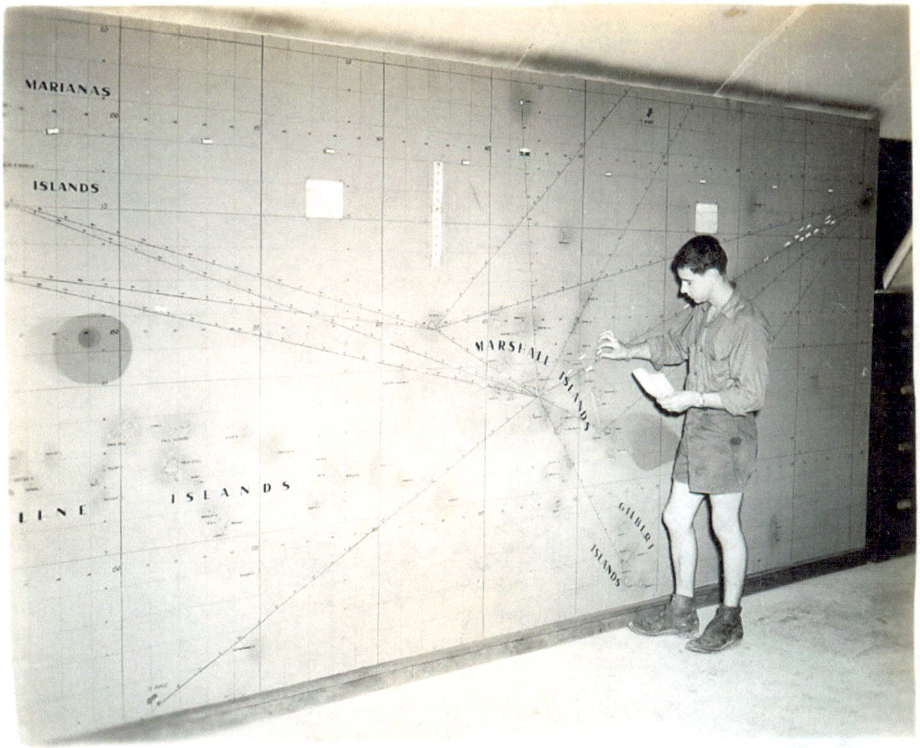

and the magnetic chart on which the position and altitude of all aircraft in this area are shown. (See HRR for Feb.)

7
Motor Pool

Between OATC and the Operations Building lies the Motor Transportation Section. Recent construction has increased garage space to four sections. The dispatcher carries on his work in the "tower" which affords excellent surveillance of the area, which is illuminated tor night control.

Part of the Motor Pool crew poses while a Corsair fighter flashes by in the background.

8
Air-Sea Rescue

On 14 May 1945 the 1535th received an OA-10 (Navy designation – PBY) for use as an air-sea rescue unit.

The crew go to work with GI brushes and elbow grease, in short order bring the plane back to a whiteness blinding under the tropical sun.

9
Transient Aircraft

In less than a month's time, the Transient Aircraft Engineering Section blossomed forth with this clean-cut area as compared with previous facilities (See photo supplement, HRR for April)—light gray building, yellow cans lettered in Red, fresh white coral sand.

SuperMechs [mechanics] for Super Fortresses [B-29s].

And now for the giant B-32 Dominator, the first of which arrived at Kwajalein on 20 May 1945.

Since October 1944, war planes have been marching out of the northeastern skies to land at Kwajalein on their flight to battle zones. Hundreds of these deadly aircraft have been serviced and thousands of their crew members fed and housed by the Air Transport Command on this island. With flaps down, a VH Baker [very heavy bomber] comes in low and level for a landing. The shore and the edge of the runway are just off to the left.

The "Follow Me" jeep [on the left side of the photo] leads the giant to her parking place, where

she joins sister ships in an orderly rank. To the far left and behind us are additional parking ramps.

The crew [of "Is This Trip Necessary?"] gets out to stretch cramped muscles and receive briefing by Assistant B-29 Engineering Officer on guarding of plane, transportation to ATC Transient area, briefing on the morrow, and loading of guns tonight. Engineering Officer checks Form 1-A and lines up work to be done on the plane.

Ammunition is distributed,

gun turrets are removed,

and the cleaning and loading of guns proceeds,

along with refueling.

The ATC B-29 Engineering Section has its offices, supplies and equipment set up in a tent area located opposite the five-plane parking ramp.

Rain and the lack of flood lighting offer slight hindrance to the work in this area. It is located just northwest of the west end of the runway. Flood lights here would blind incoming planes.

The mechanic is dwarfed by the size of the 2200 HP engine which has its air-intake dust plug [behind the bottom propeller] in place during servicing.

On the following morning, B-29 crews assemble at the line, and briefing takes place

in the Navigation and Briefing Shack. Here excellent displays of maps, charts, and blow-ups take precedence over the ever-important problem of ventilation. “Hot” pilots are the rule.

Plane serviced and crew rested and indoctrinated, another B-29 takes off to follow its destiny.

10
ISCOM Installations

Some coverage of other organizations on Kwajalein has already been submitted. Presented here are more shots of things which effect all island personnel one way or another. Island Headquarters [ISCOM or Island Command] faces the white-capped Pacific. Before it are the Flag (at half-mast for President Roosevelt) and a Jap field piece.

The Island Chapel rests beside HQ. Its open sides and louvers let the trade winds into

the quiet well-furnished interior.

As the Chapel provides food for the soul, so does the Bakery provide food for the body.

Bakers work night and day to provide bread for the island complement.

and [water] distillation units thunder constantly

at water points [storage and processing facilities] to quench the island's thirst.

Together with other theaters, the Richardson provides a place for entertainment through movies and stage shows.

In its right wing is located the Radio Station WXLG,

Purveyor of excellent stateside rebroadcasts, direct newscasts, and the inimitable Rock-Happy Roger's "Kwaj Lodge".

Baseball at Ross Field draws howling crowds.

To maintain communication with neighboring islands the U.S. Navy operates a Boat Pool that is

Well equipped with LCMs (Landing Craft Mechanized), LCVPs (Landing Craft, Vehicle, Personnel), picket boats [those used for sentinel duty], barges, and other small naval units.

11
Defense from Attack

The island of Kwajalein, although no longer considered a forward area, lies among islands and atolls of the Marshall group which still contain from 10,000 to 12,000 very-much-alive Japanese troops. In addition to these, there is the remote possibility of sneak air attack. To harass the Japanese, particularly at Wotje, as well as to maintain adequate air defense, the 4th Marine Air Wing provides F4U Corsairs and PBJs [US Navy variant of the B-25 Mitchell bomber] together with several SBDs [Douglas Dauntless], SB2Cs [Curtiss Helldiver], and sundry other tactical planes.

Radar plays its part in the detection and identification of all nearby aircraft.

and stands perpetual guard over the atoll.

Half-tracks with quadruple 50 cal. machine gun mounts stand ready for action,

as do 40mm and 90mm dual purpose guns.

On 21 March [1945] the permanent personnel of the 1535th moved from their old area (center), releasing these barracks and facilities for expanded transient traffic.

to the 422nd Sub-Depot area, west along Lagoon Road, located near the lagoon beach just to the left of the tower on the point.

Here were EM [enlisted men] barracks,

Consolidated Mess Hall,

officer's barracks,

and beach houses.

Across the road a warehouse was made available

in which was located the Orderly Room [for general administration],

recreation room and library,

Personnel Services office [notice the attractive pictures on the wall],

and Squadron Supply.

A small beer hall was set up for the EM [enlisted men],

And an addition was planned to the Sub-Depot Officer's Club to accommodate the 1535th officers.

The Raymond Clapper Memorial Theater, named for the famous newsman who lost his life in this area last year, offered 35mm movies nightly.

12
Quartering VIPs

In June, work was completed on two installations which provide more adequate quartering facilities for distinguished guests. In the ATC Transient Area, a 100’ x 16’ two story barracks was remodeled with interior partitions,

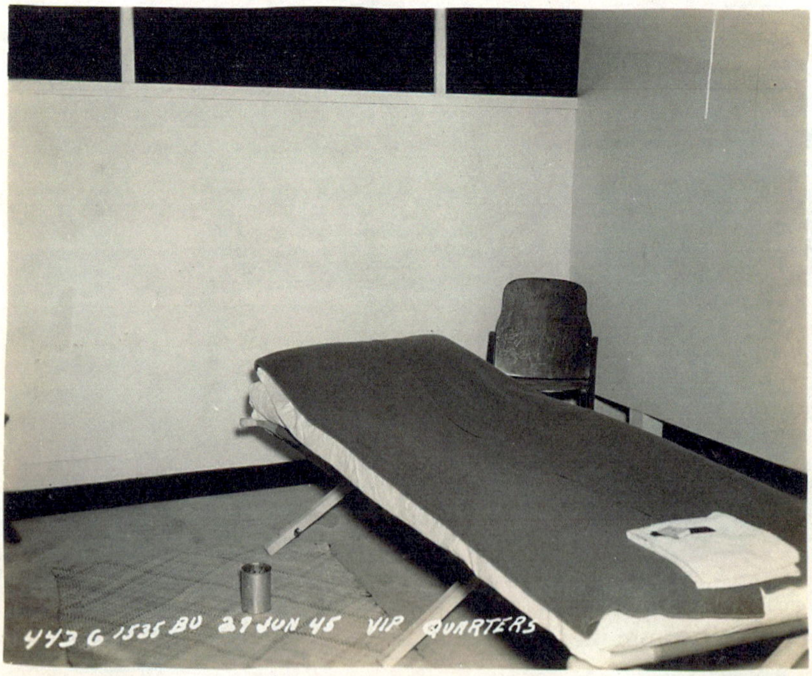

Providing light airy cubicles which are a considerable improvement over previously undivided barracks space.

In "Officer's Country" at temporary ATC quarters area, were completed additional special quarters of a comfort and luxury heretofore not encountered on the island.

Two louvered sleeping huts contain full-sized beds with innerspring mattresses.

The "VIP Lounge",

finished in cool colors and furnished for relaxation, provides a living-room,

Dining-room, and bar, with full kitchen facilities behind the shapely door at the left.

13
Free Time

In addition to baseball, shell-hunting, and stage entertainment, basketball holds its share of enthusiasts, and tournament schedules bring out expert ball-handlers.

The lagoon is a haven from furnace-heat.

Home-made and imported equipment provide gymnasts with means for not only keeping in shape, but also

for developing excellent physiques.

14
Naval Air Transport Service

Performing a mission for the Navy identical to that of the Air Transport Command but on a somewhat smaller scale, NATS Facilities are superior to those of ATC. With lumber and Quonset huts available, the entire development was planned in a minimum of time and quickly and expertly constructed by CBs [Construction Battalion].

To the rear of the line buildings, across Lagoon Road, is the NATS Hotel, separated from the ATC Transient Area by the Ready Tank area. All NATS Hotel buildings are painted battleship gray, neatly and cleanly finished inside and out.

Backing up to the beach is the Naval Officers' Club

with bright airy interior.

VIP quarters contain bedrooms, flush toilets, running water, and

a lounge with its own refrigerator, writing tables and chairs.

Officers' barracks are partitioned [inside] into two-person cubicles.

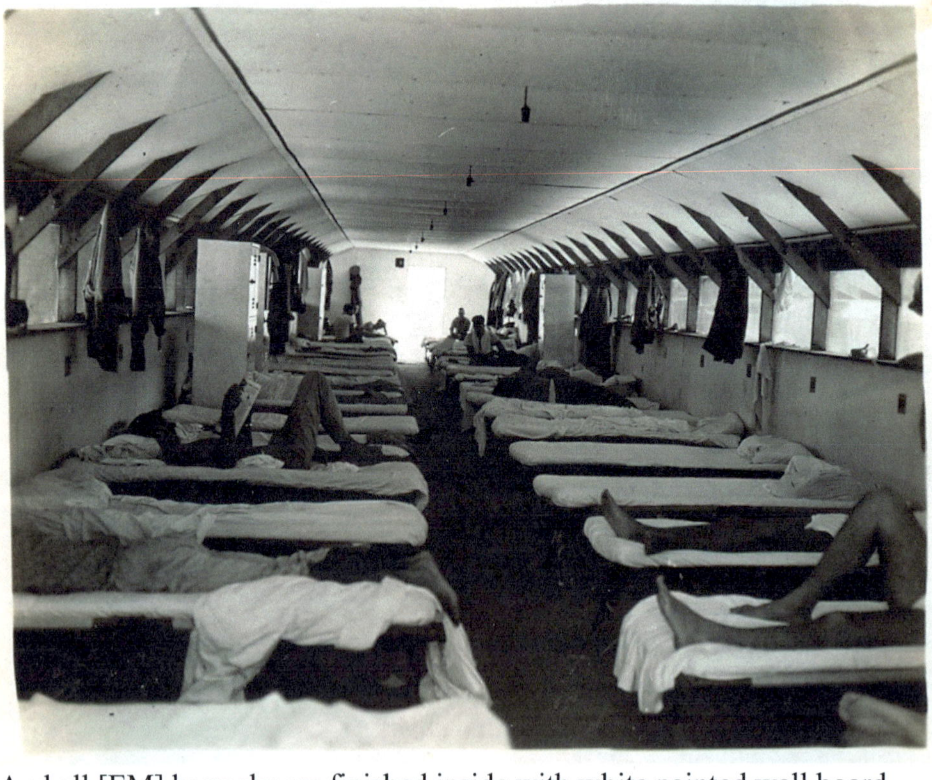

And all [EM] barracks are finished inside with white painted wall board.

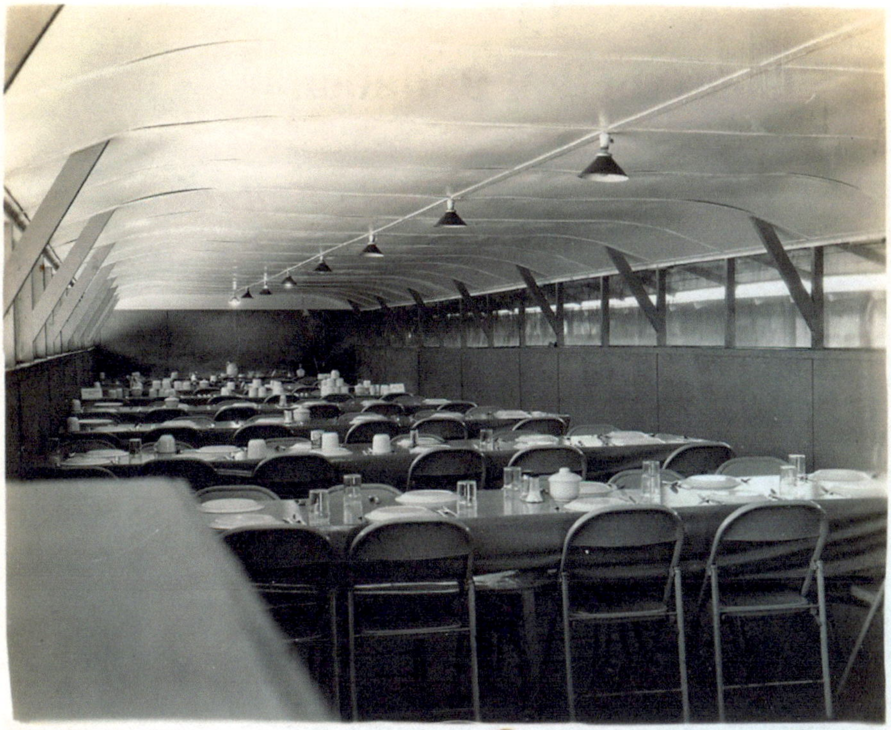

Glassware and chinaware decorate the oil-cloth-covered tables in the mess halls.

Showers are constructed for maximum ventilation and contain standard enamel wash bowls. A comparison with ATC facilities may be made by referring to HRR for March.

15
Construction Battalion

Since its arrival on 25 May, the 141st Naval Construction Battalion has pursued its mission with its usual technique of lightning efficiency. Here are highlights on its equipment and method.

The 360-ton Monighan (The House that Walks Like a Man, HRR May) plunges its 9-cubic-yard dragline shovel into the lagoon to scoop out a mountain of creamy coral.

which is trucked up a nearby ramp and dumped into a stone crusher,

consisting of several units which crush and screen the coral to various sizes and place them in separate stockpiles.

16
Concrete Mix

For the manufacture of concrete, coral of the right consistency is put up in bins by a clam shovel. From the bin it is drawn into measuring buckets, poured in an inclined conveyer where bags of cement are added. The mixture is conveyed up to the mixer skip which, when full, is raised, funneling the coral into the concrete mixer. The mixed concrete is dumped directly into trucks and driven to its destination.

17
Asphalt Plant

For surfacing roads, runways, and ramps, asphalt is manufactured in a portable plant complete in all details and devised to put out a continuous large-scale production. From coral stockpiles the aggregate is bucket-conveyed to No. 1 Stone Dryer, a heated revolving drum which extracts moisture from the mixture, and from there to No. 2 Stone Dryer which completes the drying process. It is then elevator-conveyed to the mixer where asphalt from a steam-heated storage tank is forced in and thoroughly mixed in batch form. The mixing completed, trucks carry the asphalt to Barber Greene [asphalt paver], which lays it on completed fill in a continuous 8 ft. strip of required thickness.

18
Other C. B. Tools

With their concrete and asphalt plants, the CBs brought a host of more commonly seen construction equipment. Cranes, whose spindly appearance belies their great strength; carry-alls to move tons of earth and coral,

heavy bulldozers,

and back-hoes for digging ditches and foundations. This one is tearing up the “front lawn” of ATC HQ., preparing foundations for an island electric plant.

The Raymond Clapper Theater almost disappears behind a wall of asphalt drums.

19
The Job

Using trucks, bulls, carry-alls, and graders, the topsoil is sheared from the land south of the warehouse area,

preparing it for 18 inches of pure coral fill to be black-topped with asphalt. The finished product will be the new ATC [aircraft] parking ramp.

Bulldozers and LeTourneaus [earthmoving machinery] fill in a water-filled low spot, and

coral is laid, graded, and rolled to billiard-table smoothness.

Concrete is poured for a building foundation.

Forms are set, and building material is delivered for beltline assembly of Quonset huts.

Here are shown the refueling unit

and the steam generator, principles in the fire on 11 May, described in earlier pages of this report.

During April the rains came again to Kwajalein,

and slightly inundated the low land between the line and the ATC Transient area. Poor natural drainage is being supplemented by pipelines for pumping.

19
Coral Dredge

On 13 February 1945, CB Dredge No. 1 arrived to provide this island with a supply of ballast and fill material needed for grading and providing for better drainage. Anchored in the lagoon just opposite the ATC Officers' Lounge, it can pump 120 cubic yards of coral sand an hour and has thrown up about 56,000 cubic yards since its arrival. Fantastic and beautiful shells attract souvenir hunters.

LeTourneaus and graders went to work on 21 May to spread a layer of coral in the low area between the ATC line and Lagoon Road, a section given to flooding under heavy rainfall. (See HRR for April, Plate E)

20
Entertainment

On 24 May Mr. Boris Karloff [third from left] visited the island with Maj Maurice Evans and company.

After visiting with patients at the "Trade Winds", he and the company served chow at the ATC-422nd Consolidated Mess [Hall].

Utilities Section erected a stage at the Raymond Clapper Memorial Theatre, and scenery was set against a background of the lagoon.

That night the play "Arsenic and Old Lace" was presented to an appreciative audience. [Karloff, playing the role of Jonathan (no relation), is being led off by police.]

21
Victory Through Air Power

(with apologies to Major de Seversky [influential advocate of strategic air power])

Since the start of the Pacific war, the problem of laundering grimy clothes has been met and conquered with easy ingenuity. Using "windmill washers", Army and Naval personnel have harnessed the trade winds for domestic duty. We present with pride the "Little Wonder", a four blader capable of 240 rpms [yielding material for paper pulp] at 10 mph indicated air speed,

the "Mother's Little Helper", complete with super-efficient eight-bladed fan and all-pine [deluxe] container,

the "Tall-Boy Special",

and a heart-warming group portrait of (reading from left to right) our "Monday Morning Express", "Washer's Delight" and last, but certainly not least, the "Kwajalein Mainliner", complete with sprocket drive, 14K [karat] zinc-plated drum, located in our new patented No-Prop-Wash position. For further details see our December 1944 catalogue…………

22
Award to a General

On 27 June with appropriate ceremonies, Brigadier General Ogden J. Ross, Island Commander, was presented the Navy God Star, in lieu of a second Legion of Merit, by Rear Admiral William K. Harrill, commander of Marshall-Gilbert Area. Troops are on line,

and are presented to the Reviewing Officer.

General Ross takes his position,

the National Anthem is played,

and Admiral Harrill presents the decoration.

Assuming his position on the reviewing stand, General Ross smiles at the cameraman.

The command “Pass in review” is given,

and before the reviewing officers,

the troops move smartly and with precision.

Officials of the Army, Navy, Marine Corps, and American Red Cross view the ceremony with attentive appreciation.

Col Cecil H. Childre
Former Commanding Officer of Air Transport Command at Kwajalein.

Col Childre, with ten years Army experience, was formerly CO at Hondo Field, Texas. An officer on the Central Pacific Wing Staff, Col Childre commanded the 1535th from 14 March to 14 April 1945.

Lt Col Arthur W. Stephenson
Commanding Officer of the 1535th AAF Base Unit

Lt Col Stephenson started his flying career as a cadet on 17 November 1917. In World War I, he flew as a pursuit pilot of the 25th Squadron, Fourth Pursuit Group, and has been closely associated with aviation ever since. He has amassed a total of 16,300 flying hours, 10,473 of these with one division of Western Airlines. Lt Col Stephenson was called back into Army service on 16 March 1942. He spent July and August of that year stationed at Natal, Brazil, where he started the ferrying and transport service through Ascension Island. Flying the South Atlantic and the Pacific until September 1942, he came to Hawaii and contributed to the establishment of Air Transport Command stations and routes in the Pacific area. Returning to the States, he took command of the 23rd Transport Group at Fairfield, California and operated this command until shortly before coming to Kwajalein. Lt Col Stephenson assumed command of the 1535th AAF Base Unit on 15 April 1945.

See "Ave et Vale" [hail and farewell], this issue.

Officers—1535th AAF Base Unit

Kneeling: Capt Hammond, OATC; Capt Anderson, Director, P&T; Capt Fellows, Asst. Director of Operations; Capt Harrison, Surgeon; Capt Meidel, Director S&S; Lt Cates, Adjutant; Lt Roth, P&T; **Lt Fletcher, I&S [the author of this tome and my Dad].**

Front Row: Lt Bank, P&T; Lt. Watts, Operations; Lt Sandvig, P&T; Major Borre, former CO; Capt Cavaness, Director P&AS; Capt Jenkins, Operations; Capt Lease, Engineering; Capt Hunter, Tech Supply; WOJG Gatton, Engineering (B-29); Capt Novak, Personnel Services, SWO Caldwell, Tech Inspector.

Back Row: Lt Pinkham, Stat Control; St. Zletz, Communications; Lt Parrott, Transportation; Lt. Machlis, P&T; Lt Ltalor, N&B; Major Lardent, Director, Operations; Lt Radcliff, engineering; Capt Triebel, Supply; Capt (then Lt) Pierre, Personnel; Lt Holloway, Engineering (B-29); Capt Field, Transient Service; Lt Ashworth, Mess.

23
Robert S. Fletcher, 1944 and 1972

Good work, Dad.

24
Kwaj Today

I have one final little story. I was a member of the Civil Air Patrol in Denver, Colorado in 2015 when I mentioned to one of the members of the Black Sheep Squadron the fact that my Dad was stationed on Kwajalein Island during World War II. He casually responded that he was born there. "No, you don't understand—he was stationed on Kwajalein **Island**—in the **Pacific**." Yes, he insisted, "I was **born** there." Huh? You have to be kidding, right? Nope! **He was born there**. His father was stationed there at the time. Good grief. Who'd a thunk it.

As I look over this manuscript and make any corrections I can find (so sorry for those that I missed), I am still moved by the incredible scope it represents. All these men and women who offered their service on a far-off land—and made a home there. For one brief shining moment, there was a kind of fellowship of common purpose that is a bit reminiscent of Chamelot. I do hope this story has lifted your spirit even a fraction of the extent that it has lifted mine to be able to pass it along. Few of us in later generations may be able to appreciate the gift they gave us—but we can try.

Acknowledgements

Acknowledgements in this instance must be brief. This basically is the work of my father. The list of those he might have acknowledged were unfortunately lost when he passed away. Undoubtedly, he would have acknowledged the men and women with whom he worked on Kwajalein. Let that be sufficient.

Below is the first page of the original document that my Dad gave to me.

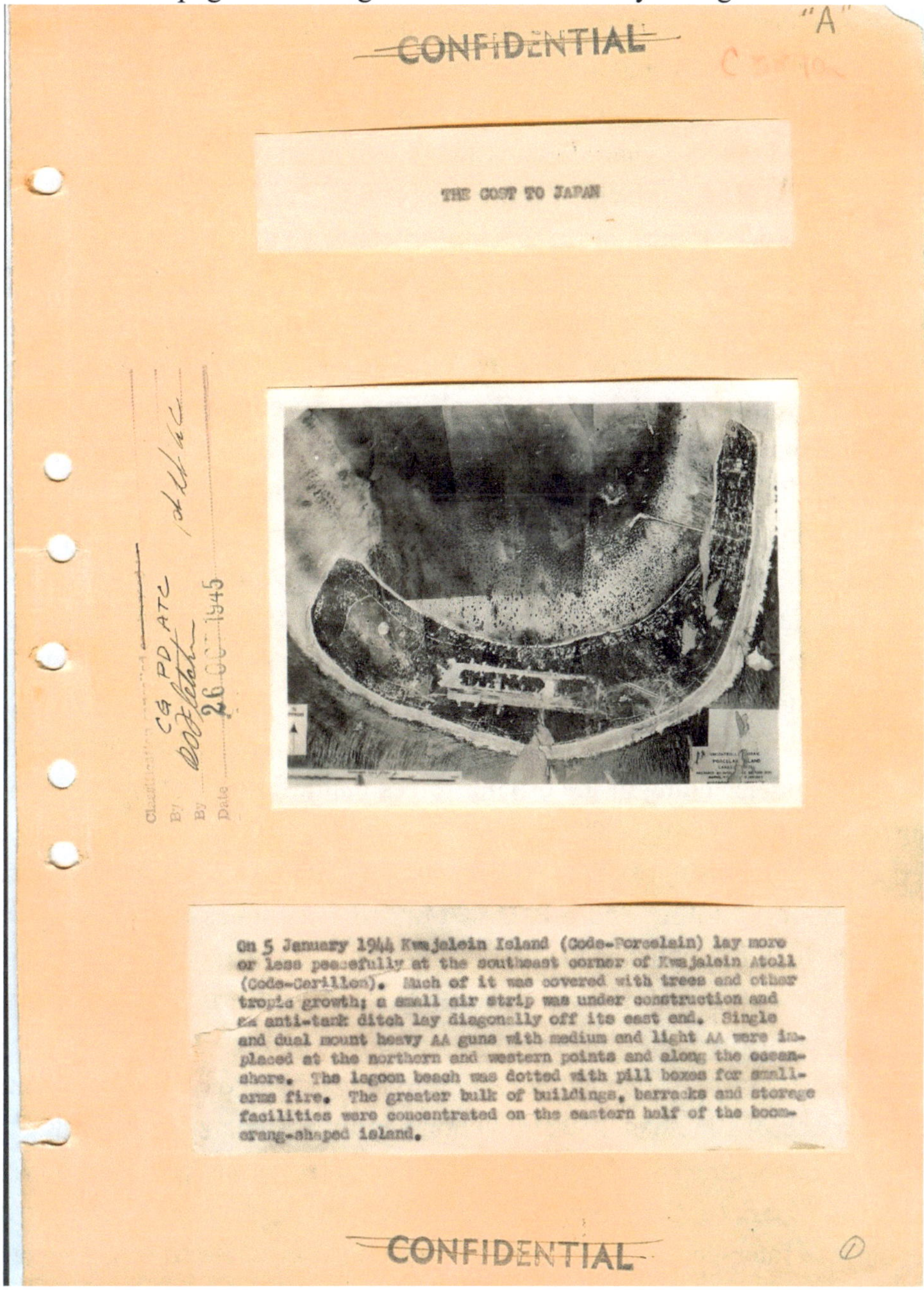

CONFIDENTIAL

"A"

THE COST TO JAPAN

Date 26 OCT 1945

On 5 January 1944 Kwajalein Island (Code-Porcelain) lay more or less peacefully at the southeast corner of Kwajalein Atoll (Code-Gerillon). Much of it was covered with trees and other tropic growth; a small air strip was under construction and an anti-tank ditch lay diagonally off its east end. Single and dual mount heavy AA guns with medium and light AA were implaced at the northern and western points and along the ocean-shore. The lagoon beach was dotted with pill boxes for small-arms fire. The greater bulk of buildings, barracks and storage facilities were concentrated on the eastern half of the boomerang-shaped island.

CONFIDENTIAL

(1)

About the Authors

Robert Fletcher was born in Portland, Maine in 1912, attended Holderness School and Bowdoin College, where he earned a BA degree in psychology. He married Gertrude Park in 1940 and enlisted in the Army Air Corps in 1944. He was discharged from the service in January 1946. and spent most of his working life after World War II at the DuPont Company. He ended his working career in the personnel department at the Savanah River Plant in Aiken, South Carolina. He died of pancreatic cancer at the age of 62 in 1974. He was a collector of guns, knives, rocks, minerals, fossils, books, and Native American artifacts. He was not an expert in anything that I can think of but was the great appreciator. I remember his attitude toward a couple of simple plastic figures he received as a result of a contribution he made to the St. Labre Indian School. I thought they were very basic—he thought they were works of art—which, of course, they were. This wondrous ability to look beyond the surface to a deeper meaning seemed to penetrate all that he did.

After we arrived in Aiken, South Carolina, Mom and Dad bought the stable of the Dolan estate and renovated it into a gracious living space. The stalls were the bedrooms, and the carriage house was converted into the kitchen, dining room, and living room downstairs and Dad's museum upstairs. Yes, it really was a museum that just scratched the surface of his broad appreciation for the human and natural worlds. To make the property a special place, Dad built a car port, a pierced-brick wall, brick walkways and central garden, and a swimming pool with the help of his amazingly versatile friend John Lamare.

Jonathan Fletcher was born in September 1946—yep, 9 months after Dad was discharged. Over the last 35 years, he has done consulting work in the area of budgeting, planning, and decision analysis in a wide range of business sectors including oil and gas, maritime transportation, and healthcare. He has developed the concept of the Integrated Planning Model and its use in Integrated Planning Systems as a means of developing a high level of managerial understanding of the driving forces that influence a firm's success as well as the areas in which risk might show itself.

He earned a BA in Chemistry from the University of the South, an MS in Geology from the University of South Carolina, and a PhD in Mineral Economics from Pennsylvania State University. In 2012 he published his first two books, The Quiz: On the Nature of the Incarnation of Jesus Christ and Human Drama Across the Curriculum: A Systems Educational Proposal. In 2014 he published his third book, The Eleventh Commandment: On the Nature of the Christian Community of Faith, followed by Into the Perfect Likeness, Scratching the Surface of Christian Transformation. Additional works include SCRAM: Avoiding Catastrophic Failure (St Maximus Scriptorium, 2015), and Models: The Art of Understanding (St Maximus Scriptorium, 2020). He has taught economics, finance, entrepreneurship, and international business at the university level. He currently resides in Hopkins, South Carolina.

Below is a picture of Mom and Dad, my sister Penny, and little not-very-old Me.

Made in the USA
Monee, IL
05 November 2026

67809035R00074